alia

 P9-EEI-803

Compare with Bears

By Kate Mineo

Gareth Stevens
Publishing

Please visit our website, www.garethstevens.com. For a free color catalog of all our high-quality books, call toll free 1-800-542-2595 or fax 1-877-542-2596.

Library of Congress Cataloging-in-Publication Data

Mineo, Kate.
Compare with bears / Kate Mineo.
    p. cm. — (Animal math)
Includes index.
ISBN 978-1-4339-5660-7 (pbk.)
ISBN 978-1-4339-5661-4 (6-pack)
ISBN 978-1-4339-5658-4 (lib. bdg.)
1. Set theory—Juvenile literature. 2. Multiple comparisons (Statistics)—Juvenile literature. 3. Bears—Juvenile literature. I. Title.
QA248.M5114 2011
511.3'22—dc22

                            2010046738

First Edition

Published in 2012 by
Gareth Stevens Publishing
111 East 14th Street, Suite 349
New York, NY 10003

Copyright © 2012 Gareth Stevens Publishing

Designer: Haley W. Harasymiw
Editor: Therese M. Shea

Photo credits: Cover, pp. 1, 5, 7 (both), 8, 9, 11 (both), 13, 15 (both), 18, 19, 21 (both) Shutterstock.com; p. 17 Philip Casey/iStockphoto.com.

All rights reserved. No part of this book may be reproduced in any form without permission in writing from the publisher, except by a reviewer.

Printed in the United States of America

CPSIA compliance information: Batch #CS11GS: For further information contact Gareth Stevens, New York, New York at 1-800-542-2595.

Contents

Boldface words appear in the glossary.

Polar Bears

Polar bears live in the **Arctic**. Polar bears are the biggest bears in the world!

Which of these polar bears is bigger?

Mother polar bears have one to four cubs at one time. The mother guards her family from enemies.

Which polar bear family has more?

7

Grizzly Bears

Grizzly bears are a kind of brown bear in North America. They have a **hump** on their shoulders.

Which picture shows more than 1 grizzly bear?

Grizzly bears eat berries, plants, animals, fish, bugs, and even dead animals! These grizzly bears are fishing.

Which picture of grizzly bears has less?

The biggest grizzlies are called Kodiak bears. They live in Alaska.

Does this Kodiak family have less than 3?

Black Bears

Black bears live in the forests of North America. Some black bears are brown, gray, or white.

Do these two groups of black bears have the same number or different numbers?

Black bears eat a lot in summer and fall so they get fat. Black bears sleep all winter in dens.

Which two black bears are the same size?

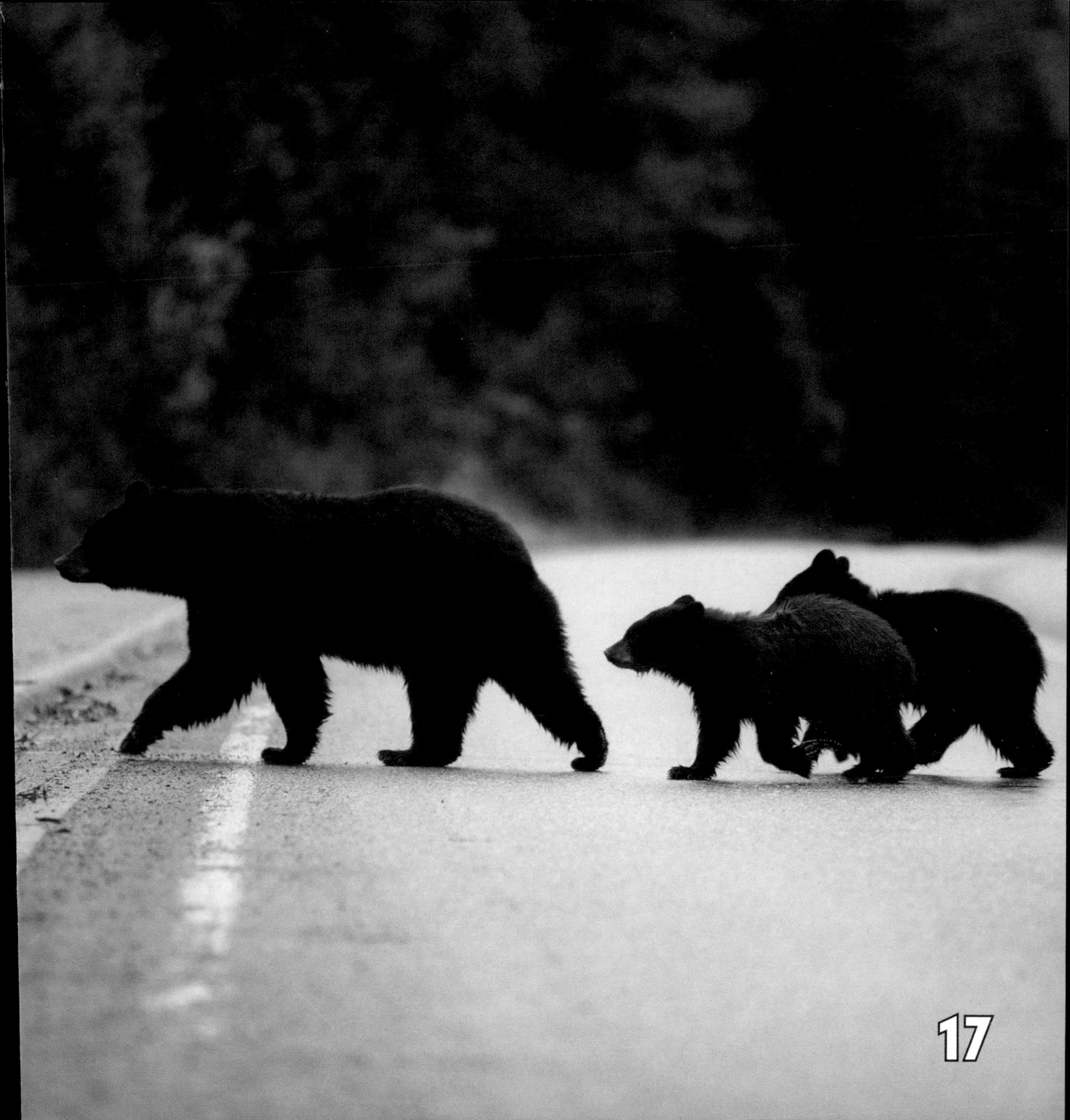

Giant Pandas

Giant pandas live in the mountains of China. They eat a grass called bamboo.

Which picture shows more pandas?

19

Giant pandas like to be alone. However, they have families. Baby pandas are very small at first.

Which panda family has more than 2?

Glossary

Arctic: the area around the North Pole

hump: a bump on an animal's back

Answer Key

For More Information

Books

Shea, Therese. *Bears*. New York, NY: PowerKids Press, 2007.

Wormell, Christopher. *Teeth, Tails, & Tentacles: An Animal Counting Book*. Philadelphia, PA: Running Press Kids, 2004.

Websites

Giant Pandas

nationalzoo.si.edu/Animals/GiantPandas/PandaFacts
Find out a lot more about pandas on the National Zoo's website.

Mammals: Polar Bear

www.sandiegozoo.org/animalbytes/t-polar_bear.html
The web page about polar bears features quick facts, photos, and a map.

Publisher's note to educators and parents: Our editors have carefully reviewed these websites to ensure that they are suitable for students. Many websites change frequently, however, and we cannot guarantee that a site's future contents will continue to meet our high standards of quality and educational value. Be advised that students should be closely supervised whenever they access the Internet.

Index